Full Auto
Volume Five

M1 Carbine to M2
Modification Manual

Desert Publications
El Dorado, AR 71731-1751

Full Auto
Volume 5
M1 Carbine to M2
Modification Manual

505 East 5th Street
El Dorado, AR 71730 U.S.A.
1-870-862-3811
info@deltapress.com

Printed in U.S.A
ISBN: 978-0-87947-130-9
10 9 8 7 6 5 4 3
Desert Publication is a division of
The DELTA GROUP, Ltd.

Direct all inquiries & orders to the above address.

Table Of Contents

WARNING

The modification of any firearm to fire fully automatic is illegal without prior approval of the BATF. Also, the manufacture of any part or group of parts that, when installed in a firearm makes it fire fully automatic, is illegal without prior BATF approval.

Please be advised that the publishing of this book is for academic purposes only. The publisher assumes no responsibility or liability for the improper or illegal modification of a firearm.

Historical Note

Preparation for war has the habit of bringing out the best in technological advances. In no case is this more true than in the development of the .30 M1 Carbine and its ammunition, the Carbine Cartridge Caliber .30 M1.

Until the 1900's, military carbines were merely adaptations of a nation's standard service rifle, featuring a shorter barrel and maybe a slightly modified stock for use by mounted cavalry. It was, of course, chambered for the same cartridge as the service rifle. The last such carbine adopted by the U.S. was the .30-40 Krag-Jorgensen prior to the Spanish-American War.

World War I changed warfare forever. Airplanes, tanks and chemical agents came into prominent use. Submarines, which had been used in a crude form even as early as the Revolutionary War, were now a sophisticated reality, a prelude to the dread U boat wolf packs of World War II. Although horses were still used as draft animals, the cavalry as such was basically a thing of the past. Transport was, so far as possible, mechanized. Machine guns wreaked their deadly havoc on all fronts and toward the end of the war, submachine guns saw service in some areas. With the Allied victory in November 1918, the "war to end all wars" was over. Advanced military thinking in the United States soundly went to sleep, with the exception of a few farsighted individuals like Billy Mitchell, who proved that battleships could be sunk by bombs dropped from airplanes.

FULL AUTO V

As war clouds gathered over Europe in the late 1930's. Americans began to cast an uneasy glance across the Atlantic. Germany and Russia had both used the Spanish Civil War as an opportunity to test new weapons and tactics, Italy had been engaged in hostilities in Ethiopia and the Japanese were gaining a vast amount of battle experience in Manchuria. When Hitler stormed Poland on September 1, 1939, the world learned a new word – *blitzkrieg*. Using this type of warfare, Germany crushed Denmark, Norway, Belgium, Luxembourg and The Netherlands in just three months in 1940. Japan had invaded China in 1937, and on September 22, 1940, Japanese troops pushed into French Indochina. The world was again at war, but for once America was pulling her pants back up even before being caught with them down at Pearl Harbor.

As early as March 1938 the U.S. Army began thinking about a light rifle or carbine as a possible replacement for the Colt .45 M1911A1 and the Thompson submachine gun. While the M1 Garand battle rifle had been adopted in 1936 as the standard service rifle, it was much too heavy and bulky to be carried by support troops who now were, thanks to aerial bombardment and other factors, in many ways more vulnerable than the front line troops doing the actual fighting. Previous practice had been to arm such troops with the Colt .45, a superior weapon for close quarters work but totally unsuited to long range shooting except by the most expert of marksmen. Even then its practical value was questionable. The initial concept for the new light rifle or carbine was for it to weigh five pounds or less, complete with sling, have an effective range of up to 300 yards, be of semi-automatic operation with capability for full automatic fire, have little noticeable recoil and be chambered for a cartridge similar to the .32 Winchester caliber for their Model 1905 autoloading rifle.

In the fall of 1940 the Ordnance Department approached Winchester with their specifications for the new cartridge. The initial cartridge was straight-walled rather than bottle-necked, with a slight taper toward the mouth to aid in ease of extraction. Its 110 gr. jacketed bullet was propelled by 14.2 grains of DuPont 4227 powder. The first production of ammunition in appreciable quantity was carried out in the spring of 1941, while weapons manufacturers were still trying to develop a suitable shoulder arm for the cartridge. In June of that year, Ordnance ordered 150,000 rounds from Winchester. A second order for 300,000 rounds followed in August.

Since Winchester was heavily involved in production of the Garand, they were unable to submit a sample design for the initial rifle tests which began on June 16. Two of the designs submitted at that time were rejected outright, and the remainder were subjected to testing with varying results. It is interesting to note that one of the samples tested and then finally rejected several months later was designed by John C. Garand, inventor of the rifle whose production had kept Winchester out of the trials! Ordnance expressed an interest in some of the samples, including Garand's, and recommended changes or modifications which might make them more serviceable. They could then be resubmitted for further testing in September. The most significant result of the initial testing was that every single design submitted was uncontrollable in fully automatic fire due to the light weight of the weapon. The full auto specification was dropped entirely.

In July Ordnance again invited Winchester to submit a sample rifle similar to an experimental design they had been working on. A mere 14 days later, a very quickly hand assembled sample was submitted to Ordnance. It was not a true indication of what the production gun would be like, as it contained numerous welds and brazes that would not be found on a production model. Yet this hastily and somewhat crudely assembled sample *worked*, and worked well. So well, in fact, that Ordnance thought it offered far more promise than any of the guns tested in June. Winchester personnel went back to work, literally day and night, and within 34 days they had a carbine ready for the September tests, finishing it just hours before the tests were to begin at Aberdeen Proving Ground. One of the principal Winchester staff members involved in this all-out effort was David Marshall Williams, universally known throughout the trade in later years simply as "Carbine" Williams. He actually designed the gas tappet which made the gun work some time before, while serving a sentence in Caledonia Prison. Williams's life story was made into a movie starring James Stewart in the early 1950 s and it's an interesting story. Keep an eye out for it on the late, late show.

The second round of tests ended on September 25, with a unanimous recommendation for the Winchester design. Ordnance awarded Winchester a contract for 350,000 of the new guns, now known as the .30 M1 Carbine, on November 24, 1941, less than two weeks before Pearl Harbor. As adopted, the ammunition for it had a slightly greater taper in the case wall and a flat rather than hollow-based bullet. Otherwise, it was essentially the same cartridge used for the intial tests.

Tooling up for large scale production is a different story than creating a prototype, however, and the first production Winchester M1's didn't come off the line until August 27, 1942. That lot consisted of a mere 10 carbines.

In addition to Winchester, a contract was awarded to the Inland Manufacturing Division of General Motors in Dayton, Ohio. By war's end, there were seven other manufacturers of the M1 Carbine: Underwood Elliott Fisher, National Postal Meter, Rock-Ola Manufacturing Corp. (a peacetime juke box manufacturer), Quality Hardware, Standard Products, Saginaw and IBM. By the end of the 1943, these nine manufacturers were turning out carbines at a rate of 500,000 per month. Winchester was, of course, the only one that was normally a firearm manufacturer. Even though they were turning out M1's on old, obsolete equipment while the other manufacturers had to tool up literally from the floor up with modern, efficient equipment, Winchester still had the lowest per unit production cost!

As the tide of war began to turn and American troops moved into new terrain and new combat situations, it became apparent that a carbine which was capable of full auto fire as in the initial specifications was desirable. Some non-military minds (most notably the Undersecretary of War) had trouble separating the concept of burst fire from the sustained fire that is normally associated with the full auto mode in machine guns. However, requests for a selective fire carbine kept coming in from those in the know, field commanders who had to watch their men die.

The basic theory was for the selective fire weapon to be used principally as a semi-auto, with short bursts of full auto only in emergencies. With very short bursts, a carbine's light weight would not be an undue handicap in controllability.

It is highly unusual for a selective fire weapon to be developed from a semi-auto only design, but that is just what happened in the case of the M2. Several methods of conversion were attempted, but the ultimate changeover was accomplished by simply redesigning a handful of working parts.

Basically, there are two types of M2 selective fire carbines. The first is the "pure" M2, which was originally manufactured as a selective fire weapon. The other is the "hybrid", starting out in life as an M1 and later converted to M2 configuration by installing a T17 kit. This conver-

The M1 Carbine was used widely in the Pacific. Note carbine on ground beside Marine artillery observers on Iwo Jima.

sion is accomplished by merely substituting parts and adding a select lever. The only other modification of the orignal weapon is the removal of some wood on the stock to accommodate the select lever. The kit as such was developed to enable the conversion of M1'a already in service without requiring that they be shipped back to an arsenal for major rework.

The M2 Carbine was approved in September 1944 and adopted as standard issue, with the M1 being relegated to a substitute standard role. In actual practice, of course, it would take some months to accomplish this. The majority of troops were still armed with M1'a at the end of the war and Germany had already surrendered when the M2 went into production. M1 assembly lines had been terminated at all plants except Winchester, Inland and Rock-Ola in 1944 as the military's need for carbines had been satisfied and production emphasis could now be placed on replacement arms. M2's went into production in May 1945 at both the Inland and Winchester plants, the only two facilities still producing carbines at that time.

By war's end over 6,000,000 M1 and M2 Carbines had been produced. Many of them would, of course, be recalled to duty in less than five years with the outbreak of war in Korea in June 1950. In American hands, they again saw some service during the Vietnam War. It is safe to say that somewhere in the world today – and tomorrow – a soldier will be staking his life on the reliability of the M1 or M2 Carbine.

From the Sands of Iwo Jima to Pork Chop Hill, the M1 Carbine and its selective fire cousin have proven themselves time and time again. They ultimately did not, of course, replace the Colt .45 M1911A1 or the Thompson entirely, but they did provide an excellent defensive carbine which also served well in offensive roles.

In recent years numerous manufacturers have produced "commercial" M1 Carbines. The best known of these firms are Universal Firearms Corporation of Hialeah, Florida and National Ordnance Incorporated of South El Monte, California. Universal's original carbine was a direct copy of the wartime model except for the handguard. Their current model is similar externally, but internally it has been changed quite a bit. Some experts feel that the current Universal M1's are actually more desirable as shooters than the real thing. National Ordnance originally began assembling finished guns in 1960 with all G.I. parts except for the barrels, receivers and a few other newly manufactured parts. Today their M1's are all newly manufactured and the parts are interchangeable with the originals.

While it is perfectly legal for anyone who can own a rifle to own an M1, ownership of the M2 is illegal without the proper Federal paperwork peculiar to automatic weapons.

The purchase or sale of an M2 involves a $200.00 transfer tax. Some M1 receivers were inadvertently stamped M2 toward the end of the war. Even though these will function only as sem-autos, the BATF still classifies them as machine guns. A partial set of parts (consisting of: 1. selector; 2. selector spring; 3. disconnector lever assembly; 4. hammer; 5. disconnector; 6. disconnector spring and 7. disconnector plunger) also constitutes a machine gun, whether they are assembled or not, even if the owner does not own a carbine to put them in. (See photograph on page 13.)

Differences Between The M1 & M2

The M1 and M2 Carbines are gas-operated, self-loading, air-cooled shoulder weapons firing the Carbine Cartridge Caliber .30 M1, contained in 15 or 30 round magazines. The M1 and M1A1 are capable of semi-automatic fire only while the M2 delivers semi-automatic or full automatic fire by means of a selector.

The M2 is essentially identical to the M1 except for the difference in design of certain components and the addition of others which permit it to deliver either semi- or full auto fire.

Component differences between the two carbines are as follows:

1. **Hammer.** The hammer of the M2 is the same as the hammer of the M1, except that it has a milled cut in the lower right side to furnish clearance for the disconnector when assembled on the hammer pin.
2. **Sear.** The sear of the M2 is the same as the sear on the M1, except for a raised shoulder on the top of the front end, which forms a camming surface for the disconnector when operated.
3. **Trigger housing.** The trigger housings of the two carbines are identical except that of the M2 has a retention slot on the left side of the magazine post

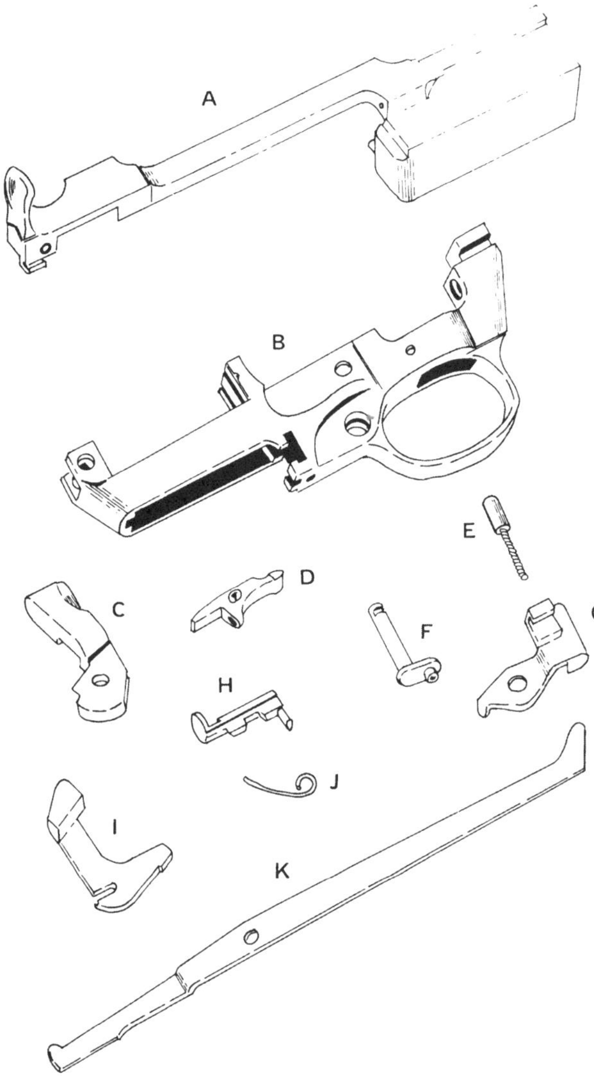

Operating parts of the M2 Carbine. A: Slide; B. Trigger housing; C. Hammer; D. Sear; E. Disconnector plunger and spring; F. Trigger housing and selector pin; G. Disconnector; H. Magazine catch; I. Selector; J. Selector Spring; K. Disconnector lever.

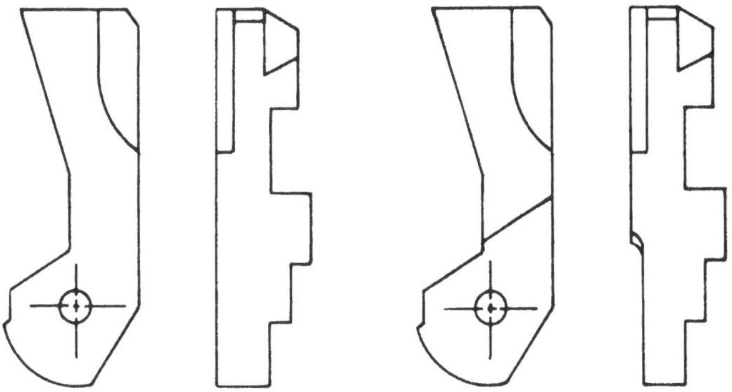

M1 hammer (left) with its M2 counterpart (right). Note milled cut in bottom of M2 hammer to furnish clearance for disconnector. (See detail drawing on page 20.)

and the front face of the post has a dismounting notch for the selector spring (the dismounting notch leads into the top of the slot). The right side has a milled cut for clearance of the disconnector lever. The left side has a milled cut for the selector.

4. **Operating slide.** The operating slide of the M2 is the same as that of the M1 except for a clearance cut extending along the right-hand side of the body, and a diagonal cut at the point where the shank of the handle joins the body. The latter cut forms a cam for camming down the forward end of the disconnector lever.
5. **Stock.** A clearance cut is made in the inner right wall of the M2 carbine to provide clearance for the projecting right side of the disconnector. A cut is also made in the inner left wall for clearance for the selector. The bridge is cut down to the central section for clearance for the disconnector lever.
6. **Disconnector group (added parts).** The disconnector pivots on the hammer pin when assembled. The rear

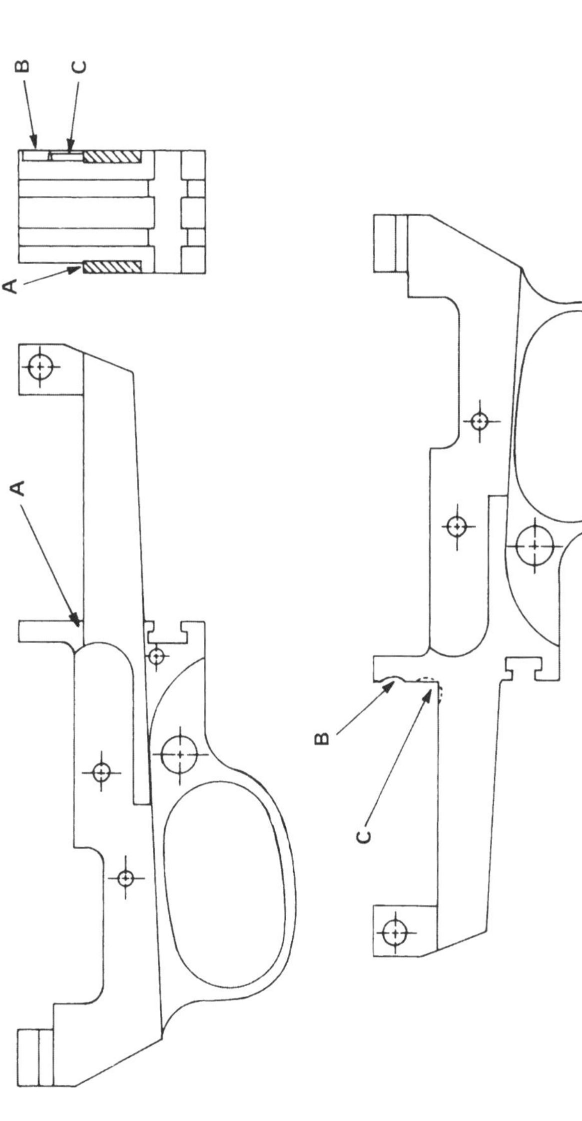

M2 trigger housing. Note milled cut for clearance of disconnector lever (A), selector spring dismounting notch (B) and selector spring retention slot (C). (See drawings on pages 21 and 22.)

Wartime M1 Carbine ready for conversion to M2 configuration. The M1 is one of the simplest weapons to convert to full auto since it involves merely substituting some parts and adding a few others.

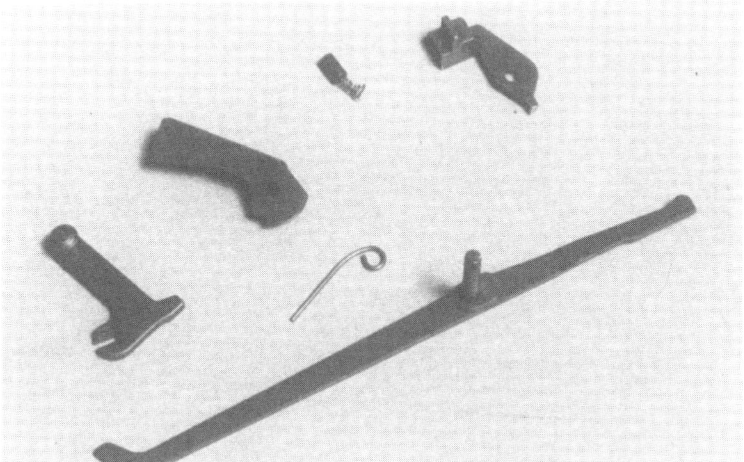

Would you believe – a machine gun? According to the BATF the 7 parts illustrated above are a machine gun, even though there is no way they in themselves could be made to fire even a single shot. The BATF ruling applies to these specific M2 parts since they are capable of making a semi-auto M1 into a full auto M2. They are, clockwise from left center: selector, hammer, disconnector plunger and spring, disconnector, disconnector lever assembly and selector spring.

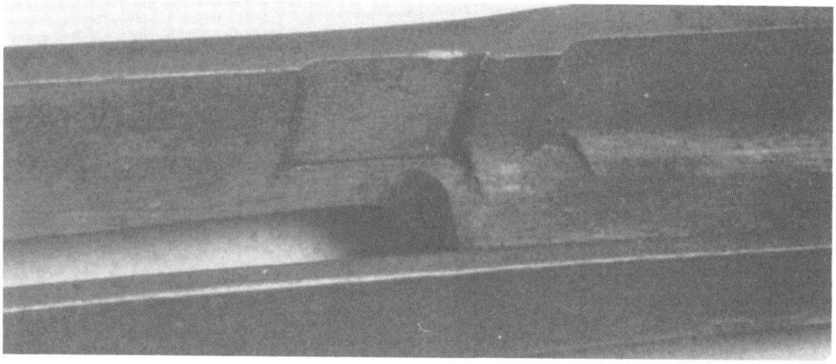

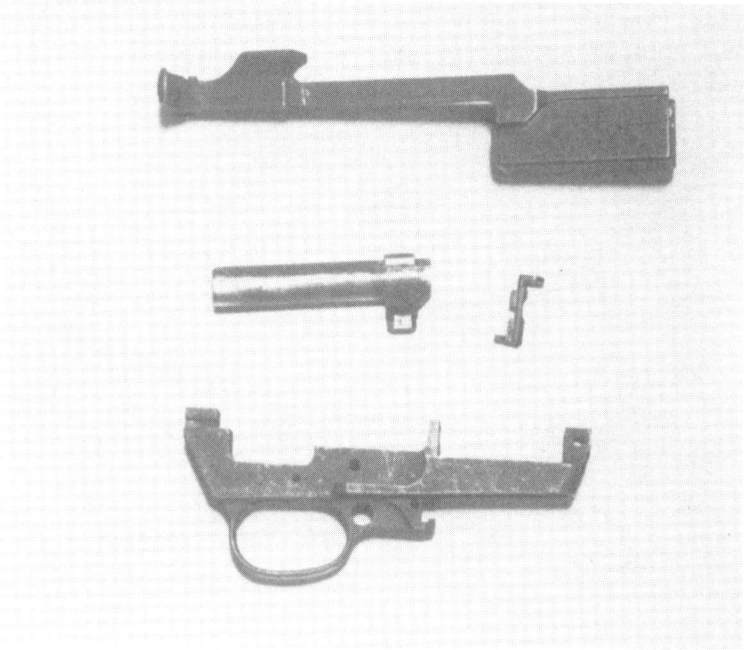

Top: Stock for M2 has a cutout for the selector lever within the inletted area on the left-hand side. In this view, the butt-stock is to the left and the forestock to the right. Bottom: M2 operating slide, bolt, magazine catch and trigger housing. These parts and the M2 stock can be used on a standard M1 without affecting its semi-auto only status. However, the corresponding M1 parts cannot be used on the M2.

M1 and M2 slides for comparison's sake. Note that the slide of the M1 at top is flat on the right side of the body while the slide for the M2 has a clearance cut along the side and a diagonal camming cut where the handle joins the body.

end has a lateral projection, which bears upon the raised shoulder of the sear, when the disconnector is cam-operated by the disconnector lever for full automatic fire. The forward end has a projecting lug on the right side, which extends outside the trigger housing and engages and acts as a camming surface for the rear end of the disconnector lever. A spring and a plunger, recessed in the top of the disconnector and bearing on the receiver, return the disconnector to the inoperative position, when the camming action of the lever is discontinued.

7. **Disconnector lever assembly (added parts).** The disconnector lever assembly should not be disassembled. If any part is found to be defective, the entire assembly should be replaced. The disconnector lever assembly is composed of a disconnector lever pin, disconnector lever rivet and disconnector lever. The pin retains the trigger housing when assembled to the receiver. The disconnector lever, riveted to a pivot on the pin, is shifted in a vertical plane, by the turning of the pin. The pin is turned by the selector. A rounded projection on the rear end of the disconnector lever operates the disconnector. A projecting toe on the front end of the disconnector lever contacts the camming surface on the operating slide. An offset in the rear section provides for alignment with the disconnector.
8. **Selector group (added parts).** The selector is mounted to the left end of the crank pin by means of a slot in the lower forward face of the selector mating the straddle slots in the end of the pin. The selector holds the pin in position and acts as a lever for turning, throwing the disconnector lever into or out of engagement with the operating slide. A curved wire spring holds the selector in position on the pin and in the fully automatic or semi-automatic position when operated. The straight front end of the spring seats into a recess in the lower rear end of the selector and the circular rear end of the spring seats in a vertical slot in the front face of the magazine post on the left side of the trigger housing. When assembled, the circular end of the spring is downward (spring is concave downward).

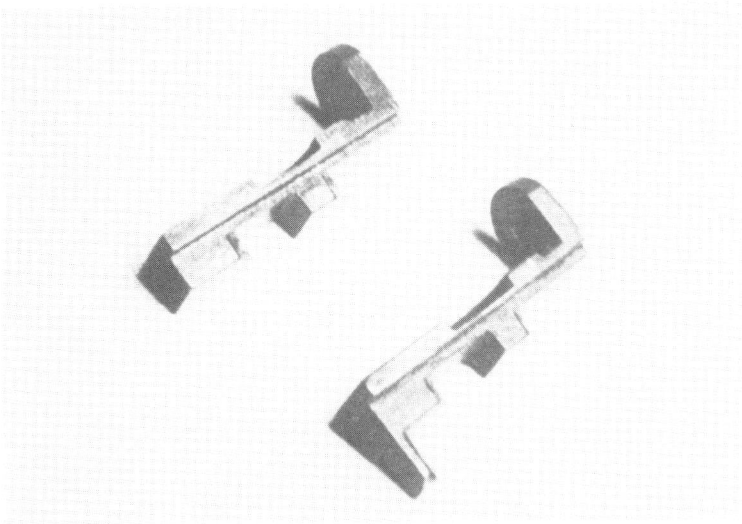

Old style magazine catch (top) for M1, new style for M2 (below). Note the extension on the bottom of the latter which can also be used on M1's.

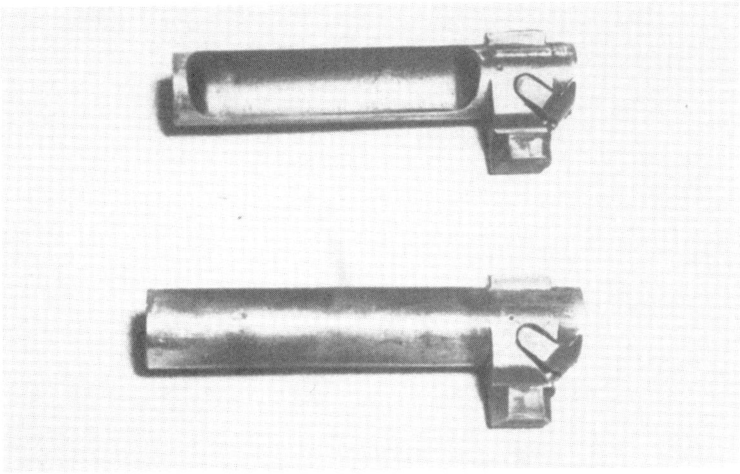

Old style M1 bolt (top) should not be used for an M2 conversion job. The later M2 style bolt (below) is the only one that should be used.

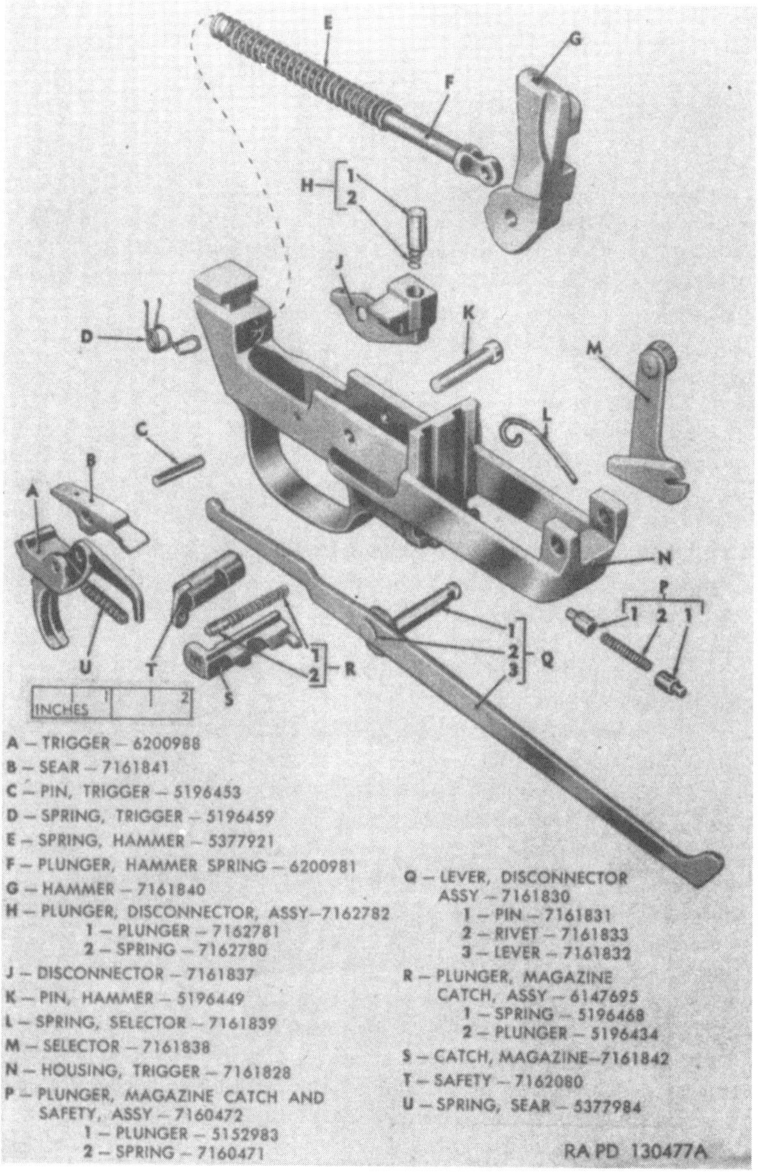

Trigger housing group for the M2 Carbine.

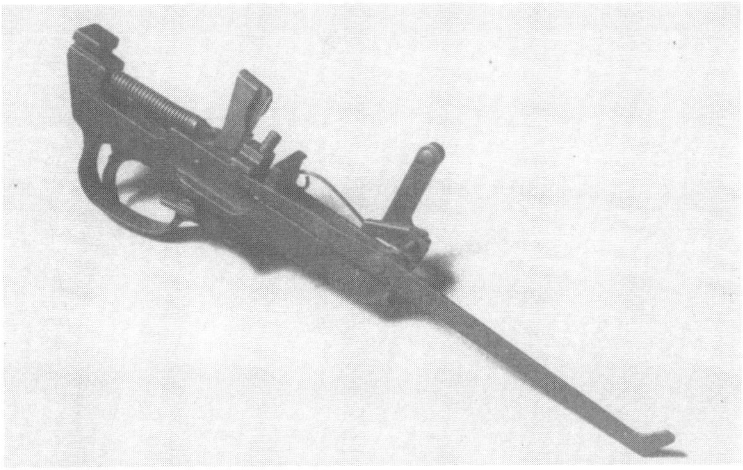

Functioning of trigger housing group.

FUNCTIONING OF THE TRIGGER HOUSING GROUP (AUTOMATIC SETTING)

When the carbine is fired automatically, the trigger housing group functions almost the same as on the semiautomatic setting. On the automatic setting, however, each time the operating slide moves forward the rear end of the disconnector lever is rotated upward causing the disconnector to pivot about the hammer pin. When the trigger is held to the rear, the rear part of the disconnector contacts the raised shoulder of the sear and forces the forward end of the sear down, disengaging the sear nose from the hammer. The hammer moves forward, actuated by the hammer spring, and the carbine fires. If the trigger is released, the sear moves to the rear under action of the hammer spring, and the rear of the disconnector cannot contact the forward end of the sear. The sear nose remains engaged with the hammer, the hammer is held to the rear, and the carbine stops firing.

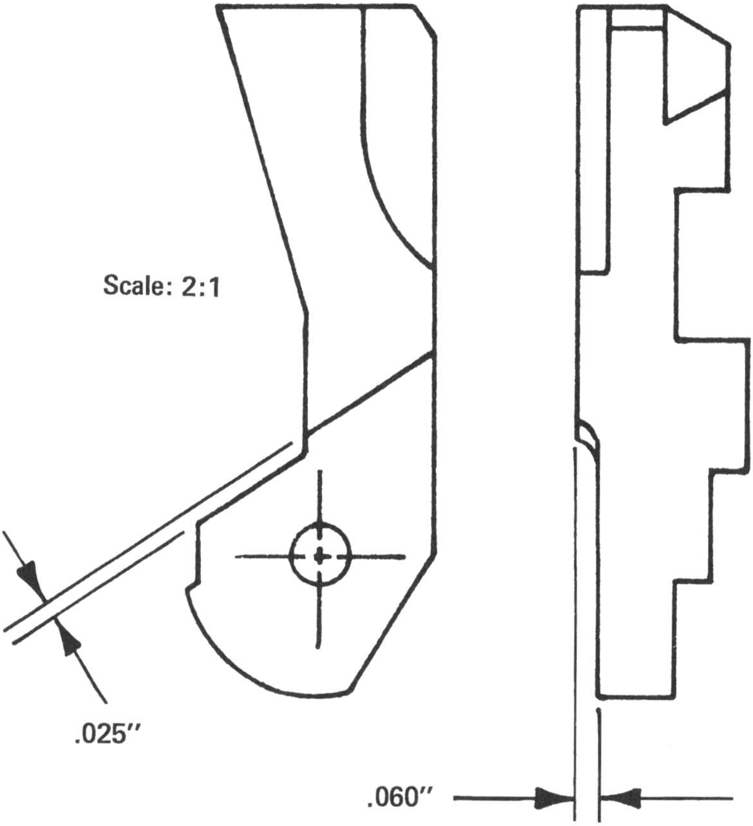

Mill cut on hammer.

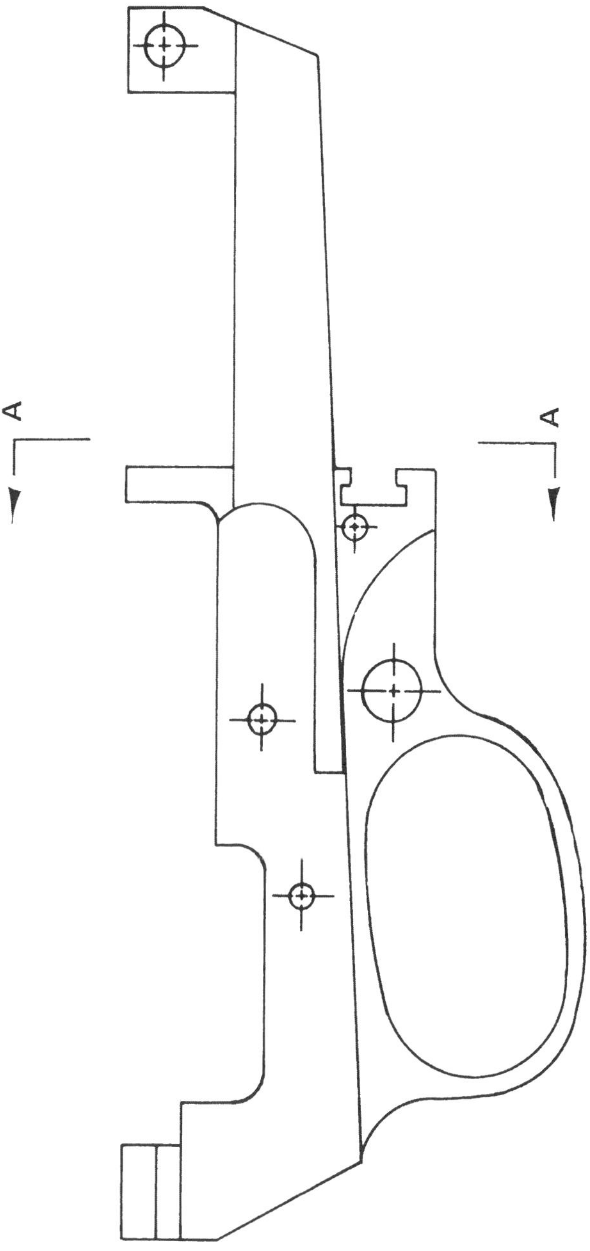

See pages 22, 23 and 24 for trigger housing machining details.

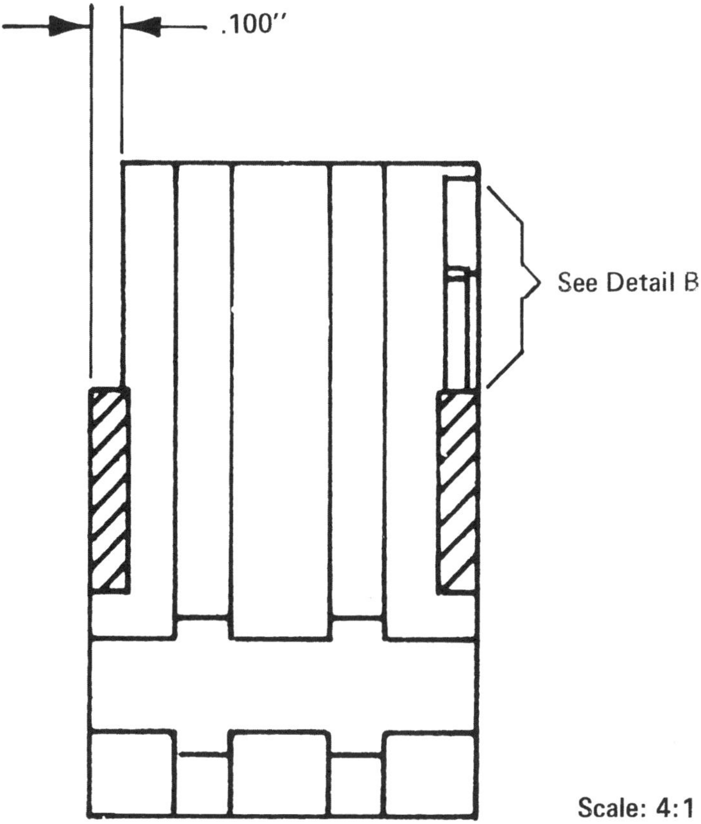

Section A-A, rotated 90° clockwise.

Scale: 2:1

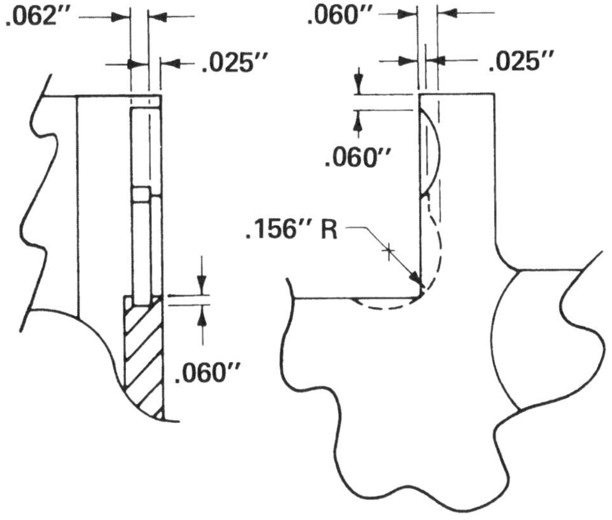

Detail B

See next page for detail of milling cutter.

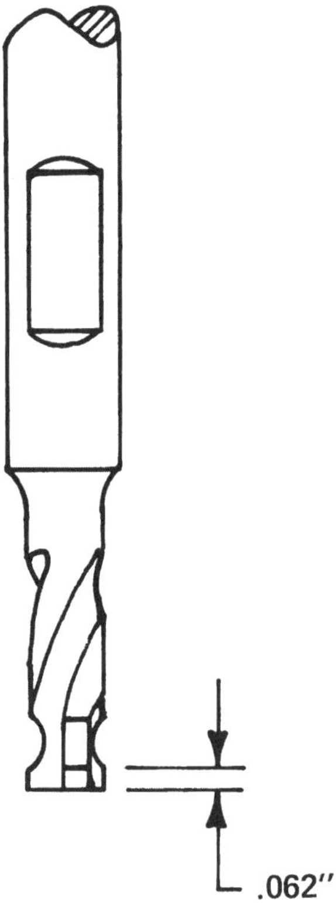

Milling cutter for selector spring slot can be made from standard 5/16″, 4 flute end mill.

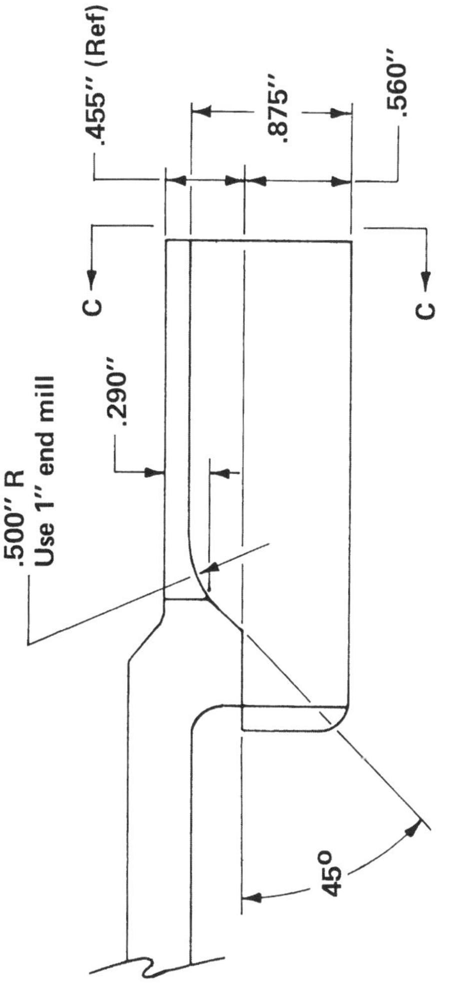

Mill cut on bolt.

Scale: 2:1

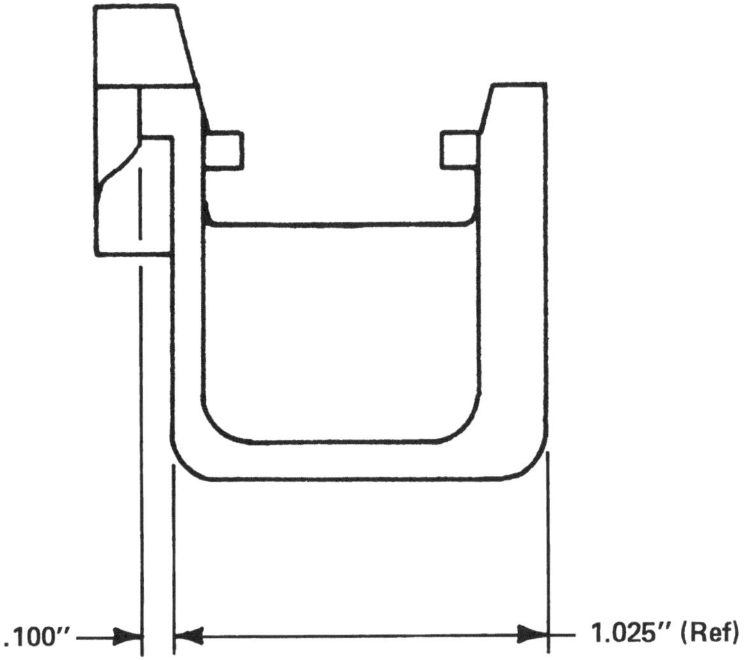

View CC

Assembled M2 less stock.

Other Books Available From Desert Publications

081	Crossbows/ From 35 Years With the Weapon	$15.95
206	CIA Field Expedient Preparation of Black Powder	$19.95
221	Evaluation of Improvised Shaped Charges	$19.95
411	Homeopathic First Aid	$11.95
454	Survival Childbirth	$15.95
C-002	How To Open A Swiss Bank Account	$9.95
C-011	Defending Your Retreat	$9.95
C-020	Methods of Long Term Storage	$9.95
C-023	Federal Firearms Laws	$11.95
C-024	Select Fire Uzi Modification Manual	$13.95
C-025	How to Build Silencers	$6.95
C-027	Firearms Silencers Volume 1	$13.95
C-029	M1 Carbine Arsenal History	$7.95
C-032	The Silencer Cookbook	$9.95
C-038	How To Build A Beer Can Morter	$4.95
C-040	Criminal Use of False ID	$14.95
C-045	Improvised Weapons of American Underground	$14.95
C-048	Catalog of Military Suppliers	$18.95
C-056	Navy Seal Manual	$29.95
C-065	Poor Man's James Bond Vol 1	$34.95
C-113	Submachine Gun Designers Handbook	$19.95
C-114	AR-15, M16 and M16A1 5.56mm Rifles	$13.95
C-117	Ruger Carbine Cookbook	$13.95
C-120	Shootout II	$14.95
C-122	AK-47 Assault Rifle	$15.95
C-123	Combat Loads for Sniper Rifles	$18.95
C-125	Browning Hi-Power Pistols	$13.95
C-126	UZI Submachine Gun	$13.95
C-127	FullAuto Vol 1 Ar-15 Modification Manual	$11.95
C-128	Full Auto Vol 2 Uzi Mod Manual	$11.95
C-130	Full Auto Vol 4 Thompson SMG	$11.95
C-131	FullAuto Vol 5 M1 Carbine to M2	$11.95
C-135	M-14 Rifle, The	$12.95
C-136	Fighting Garand Manual	$13.95
C-137	Ranger Training & Operations	$19.95
C-138	Emergency War Surgery	$29.95
C-140	Emergency Medical Care/Disaster	$14.95
C-143	Survival Medicine	$15.95
C-150	Improvised Munitions Black Book Vol 1	$17.95
C-151	Improvised Munitions Black Book Vol 2	$17.95
C-152	Impro. Munitions Black Book Vol 3	$29.95
C-155	Fighting Back on the Job	$14.95
C-162	Colt .45 Auto Pistol Manual	$13.95
C-163	Survival Evasion & Escape	$19.95
C-165	USMC Sniping	$19.95
C-175	Beat the Box	$7.95
C-197	How to Make Disposable Silencers Vol. 1	$19.95
C-198	Expedient Hand Grenades	$21.95
C-199	U. S. Army Counterterrorism Training Manual	$19.95
C-203	Explosives and Propellants	$19.95
C-204	Improvised Munitions/Ammonium Nitrate	$19.95
C-205	Two Component High Explosives Mixtures	$19.95
C-207	CIA Improvised Sabotage Devices	$19.95
C-208	Spcial Forces Demolitions Training Handbook	$19.95
C-210	Guerilla Warfare	$17.95
C-211	FN-FAL Auto Rifles	$21.95
C-218	Professional Homemade Cherry Bomb	$15.95
C-219	Improvised Shaped Charges	$19.95
C-220	Improvised Explosives, Use In Detonation Devices	$19.95
C-227	Springfield '03 Rifle Manual	$15.95
C-229	Lock Picking Simplified	$11.95
C-230	How to Fit Keys by Impressioning	$9.95
C-231	Lockout -Techniques of Forced Entr	$15.95
C-235	How to Open Handcuffs Without Keys	$13.95
C-244	Camouflage	$14.95
C-274	M14 and M14A1Rifles and Rifle Marksmanship	$23.95
C-284	Thompson Submachine Guns	$19.95
C-289	M1 Carbine Owners Manual	$13.95
C-290	Clandestine Ops Man/Central America	$14.95
C-324	How to Make Disposable Silencers Vol. 2	$19.95
C-333	USMC AR-15/M-16 A2 Manual	$21.95
C-337	Sten MKII SMG Construction Manual	$24.95
C-356	Map Reading and Land Navigation	$19.95
C-365	Survival Gunsmithing	$14.95
C-372	Poor Man's James Bond Vol 2	$34.95
C-386	Self-Defense Requires No Apology	$15.95
C-414	Select Fire 10/22	$15.95
C-415	Construction Secret Hiding Places	$15.95
C-416	Brown's Alcohol Motor Fuel Cookbook	$24.95
C-417	Canteen Cup Cookery	$9.95
C-418	Improvised Batteries/Detonation Devices	$19.95
C-421	Walther P-38 Pistol Manual	$13.95
C-422	P-08 Parabellum Luger Auto Pistol	$14.95
C-423	Sten Submachine Gun, The	$11.95
C-427	How to Train a Guard Dog	$16.95
C-429	The Anarchist Handbook Vol. 1	$14.95
C-430	Beretta - 9MM M9	$15.95
C-432	Mercenary Operations Manual	$14.95
C-439	Improvised Lock Picks	$15.95
C-455	Modern Day Ninjutsu	$17.95
C-512	FullAuto Vol 8 M14A1 & Mini 14	$11.95
C-518	L.A.W. Rocket System	$14.95
C-528	With British Snipers, To the Reich	$29.95
C-539	Dental Specialist	$59.95
C-558	Food, Fur and Survival	$17.95
C-561	The Squeaky Wheel	$17.95
C-562	Survival Chemist Book	$17.95
C-564	Medical Specialist	$49.95
C-571	U. S. Marine Corps Scout/Sniper Training Manual	$27.95
C-610	Aunt Bessie's Wood Stove Cookbook	$9.95
C-622	Defensive Shotgun	$14.95
C-625	MAC-10 Cookbook	$14.95
C-627	Brown's Lawsuit Cookbook	$26.95
C-628	Can You Survive	$16.95
C-633	Hand to Hand Combat	$11.95
C-634	USMC Hand to Hand Combat	$9.95
C-635	Hand to Hand Combat by D'Eliscue	$11.95
C-636	Prisons Bloody Iron	$16.95
C-639	Police Karate	$17.95
C-646	Science of Revolutionary Warfare	$16.95
C-668	Survival Shooting for Women	$7.48
C-680	Infantry Scouting, Patrol. & Sniping	$16.95
C-694	CIA Field Expedient Incendiary Manual	$19.95
C-761	How to Lose Your X-Wife Forever	$24.95
C-763	Vigilante Handbook	$19.95
C-775	Alcohol Distillers Handbook	$27.95
C-776	How to Build a Junkyard Still	$15.95
C-777	Engineer Explosives of WWI	$19.95
C-794	US Marine Corps Essential Subjects	$19.95
C-815	US Marine Bayonet Training	$11.95
C-818	The Poisoner's Handbook	$29.95
C-823	Cold Weather Survival	$14.95
C-829	The Anachist Handbook Vol. 2	$14.95
C-889	Napoleon's Maxims of War	$11.95
C-890	Company Officers HB of Ger. Army	$19.95
C-891	German Infantry Weapons Vol 1	$18.95
C-896	Op. Man. 7.62mm M24 Sniper Weapon	$13.95
C-899	US Army Bayonet Training	$11.95
C-9002	German MG-34 Machinegun Manual	$11.95
C-9012	Firearm Silencers Vol 3	$18.95
C-9013	HK Assault Rifle Systems	$27.95
C-9047	SKS Type of Carbines, The	$17.95
C-9048	Rough Riders, The	$24.95
C-9052	Private Weaponeer, The	$17.95
C-9057	Keys To Understanding Tubular Locks	$11.95
C-9060	The Anarchist Handbook Vol. 3	$14.95
C-9080	Leadership Hanbook of Small Unit Ops	$14.95
C-9081	Dirty Fighting	$13.95
C-9082	Lasers & Night Vision Devices	$29.95
C-9083	Water Survival Training	$8.95
C-9084	Assorted Nasties	$29.95
C-9085	Ruger P-85 Family of Handguns	$19.95
C-9101	Guide to Germ Warfare	$17.95
C-9102	CIA Field Expedient Method for Explosives Preparation	$19.95
C-9119	Surviving Global Slavery	$16.95
C-9120	Military Knife Fighting	$14.95
C-9127	H&R Reising Submachine Gun Manual	$19.95
C-9131	Live to Spend It	$29.95
C-9134	Military Ground Rappelling Techniques	$15.95
C-9135	Smith & Wesson Autos	$19.95
C-9137	Caching Techniques of U.S. Army Special Forces	$14.95
C-9138	USMC Battle Skills Training Manual	$49.95
C-9139	Survival Armory	$27.95
C-9156	Concealed Carry Made Easy	$17.95
C-9164	The L'il M-1, The .30 Cal. M-1 Carbine	$17.95
C-9170	Urban Combat	$27.95
C-9178	Apocalypse Tomorrow	$17.95
C-9198	MP40 Machinegun Manual	$16.95
C-9199	Clear Your Record & Own a Gun	$24.95
C-9200	Sig's Handguns Manual	$19.95
C-9210	Sniper Training	$24.95
C-9212	Poor Man's Sniper Rifle	$19.95
C-9221	The Official Makarov Pistol Manual	$14.95
C-9224	The Butane Lighter Hand Grenade	$14.95
C-9228	Unarmed Against the Knife	$14.95
C-9229	Black Book of Booby Traps	$29.95
C-9239	Militia Battle Manual	$22.95
C-9247	Black Book of Revenge	$15.95
C-9253	Bankruptcy, Credit and You	$18.95
C-9255	The Sicilian Blade	$15.95
C-9259	How to Build Practical Firearm Silencers	$18.95
C-9262	Glock's Handguns	$19.95
C-9263	Heckler and Koch's Handguns	$19.95
C-9264	The Poor Man's Ray Gun	$11.95
C-9265	The Poor Man's R. P. G.	$24.95
C-9279	Credit Improvement & Protection Handbook	$19.95
C-9281	Build Your Own AR-15	$19.95
C-9282	How to Live Safely in a Dangerous World	$19.95
C-9293	Bllack Book of Arson	$24.95
C-9322	How to Build Flash/Stun Grenades	$19.95
C-9342	Entrapment	$19.95
C-9359	The FN-FAL Rifle Et Al.	$21.95
C-9367	The Mental Edge, Revised	$17.95
C-9373	Survival Bible by Duncan Long	$59.95
C-9405	The Internet Predator	$17.95
C-9406	The Criminal Defendants Bible	$49.95
C-9430	Protect Your Assets	$24.95
C-9432	How To Build Military Grade Suppressors	$21.95
C-9441	Mossberg Shotguns	$24.95
C-9478	Poor Man's Primer Manual, The	$24.95
C-9546	Agents HB of Black Bag Ops	$14.95
C-9556	Improvised Rocket Motors	$14.95
C-9612	Homestead Carpentry	$14.95

PRICES SUBJECT TO CHANGE WITHOUT NOTICE

870-862-3811

Desert Publications
505 East Fifth St., Dept. - BK - 130
El Dorado, AR 71730
www.deltapress.com

Shipping & Handling
1 item $7.95 - 2 or more
$11.95